NASA's Artemis III Mission

A Journey Into the Lunar Unknown And Everything You Need To Know

James J. Smith

All rights reserved. No part of this publication may be reproduced, distributed, or transmitted in any form or by any means, including photocopying, recording, or other electronic or mechanical methods, without the prior written permission of the publisher, except in the case of brief quotations embodied in critical reviews and certain other noncommercial uses permitted by copyright law.

Copyright © **James J. Smith** 2024

Table of contents

Chapter 1
Introduction to Artemis III Mission
Overview of Artemis 3
Mission Objectives and Timeline
Chapter 2
Mission Planning and Development
Crew Selection and Training
Spacecraft Overview: Orion and Starship HLS
Chapter 3
Launch and Transit
Space Launch System (SLS) Overview
Launch Site and Vehicle
Journey to Lunar Orbit
Fine-Tuning the Trajectory
Chapter 4
Lunar Exploration
Goals of Artemis 3
Landing Site Selection
Surface Activities and Scientific Objectives
Chapter 5
Crew Operations and Equipment
Scientific Instruments and Discoveries
Challenges and Delays
Conclusion

Intentionally left blank

Chapter 1

Introduction to Artemis III Mission

Visualize a gigantic white rocket, towering over the Statue of Liberty, exploding in a spectacular display of fire as it penetrates the atmosphere. On board, a crew of space travelers, the product of years of intense preparation and unfaltering resolve, race towards a celestial object that has not been visited by humans in more than fifty years. The Artemis III mission, a monumental undertaking that will change the course of history and reshape humanity's relationship with the Moon, is here and now, not in some faraway science fantasy realm.

With "one small step for a man, one giant leap for mankind" onto the lunar surface, the world watched in awe fifty years ago, when Neil Armstrong stepped onto the moon. A sense of awe and the shared goal of space exploration were kindled by the Apollo missions. Unfortunately, that ambition lay dormant for a long time. With Artemis III, humanity is prepared to rekindle the fire and begin a new era of lunar exploration, one that is both more ambitious and revolutionary than the last.

This isn't just Apollo all over again. The scientific community is behind the Artemis III mission, which aims to further human exploration. Its destination: the intriguing South Polar Region of the Moon, a realm of eternal shadows and buried secrets. Here, tucked beneath the lunar surface, lies the possibility for a treasure trove of knowledge – frozen water ice, a resource necessary for establishing a sustainable human presence on the Moon. This ice might be used for drinking water, fuel generation, and even radiation protection for future lunar settlements. Unraveling the secrets of the south pole could be the key to unlocking the Moon's potential as a stepping stone towards Mars and beyond.

But the ultimate captivation of Artemis III resides in its crew. This mission will see a historic first — the landing of the first woman and the first person of color on the lunar surface. This is a significant moment, not only for space exploration, but for mankind as a whole. It symbolizes a significant stride towards inclusivity and smashing glass ceilings in the once-exclusive field of spaceflight. The narrative of these astronauts, their devotion and the hurdles they overcame, will be an inspiration for decades to come.

The voyage to Artemis III is no simple feat. It involves a complex ballet of perfectly constructed spacecraft working in perfect unison. The gigantic Space Launch System (SLS) rocket, the most powerful ever created, will blast the Orion spacecraft, humanity's new lunar chariot, out of Earth's hold. Aboard Orion, the astronauts will embark on a multi-day voyage, their home away from home across the wide expanse of space.

But Orion is just one component of this complicated puzzle. Await them on the lunar surface is the Starship HLS, a groundbreaking reusable lander created by SpaceX. This future transport will gently touch down on the lunar surface, delivering the astronauts to their exploring destination. Imagine the awe-inspiring image of the Starship HLS soaring back into the lunar sky after a successful mission, taking the astronauts back to Orion for their return trip to Earth.

The intended lunar stay is an adventure in itself. Clad in next-generation spacesuits, the astronauts will walk out onto the alien surface, conducting a variety of scientific experiments and gathering vital lunar samples. They will explore craters steeped in constant darkness, collect data on the lunar environment, and deploy cutting-edge scientific instruments. These equipment will research the Moon's internal structure, assess the makeup of the lunar soil, and even investigate the possibilities for cultivating crops in the hard lunar environment.

Each finding, each new morsel of information, will add to a deeper understanding of our lunar neighbor and pave the way for future research.

Artemis III is much more than a landing mission; it's a scientific odyssey. It's a testimony to human intellect, a culmination of decades of technological achievement. It's a narrative of daring and perseverance, of pushing the boundaries of human exploration and aiming for the seemingly impossible. It's a promise of a future where humanity has a permanent footprint on the Moon, where the lunar surface is no longer a desolate wasteland, but a lively hub of scientific inquiry.

This book, "NASA's Artemis III Mission: A Journey Into the Lunar Unknown," is your invitation to be a part of this historic expedition. We will delve into the details of the mission, the science behind it, and the stories of the brave astronauts who will make the voyage. We will explore the technical marvels of the spacecraft, the obstacles that lie ahead, and the great possibilities this mission has for the future of space travel.

So, buckle up, my reader. Prepare to have your imagination ignited and your perspective on humanity's place in the universe forever altered. This is the story of Artemis III, and the trip begins here.

Overview of Artemis 3

Artemis III, planned for launch no sooner than September 2026, symbolizes a tremendous leap forward in human space exploration. It's the first crewed mission of the Artemis program, aiming to land the first woman and person of color on the Moon's South Pole region. This dangerous endeavor isn't just about planting a flag; it's a scientific expedition paving the way for a continuous human presence on the Moon.

Destination: The Lunar South Pole

Unlike the Apollo missions that landed around the equator, Artemis III sets its sights on the Moon's South Pole. This hitherto undiscovered region has immense scientific intrigue. Here, persistently shadowed craters store potential reservoirs of water ice, a critical resource for future lunar communities. The ice might be utilized for drinking water, manufacturing rocket fuel, and even providing radiation shielding. Studying the nature and distribution of this ice will be a primary priority for the Artemis III crew.

A Crew for the Future

Artemis III will carry a crew of four astronauts, two of whom will be picked for the historic lunar landing. This mission marks an important milestone for diversity and inclusiveness in space exploration. The crew will undertake intense preparation, preparing them for the physical and psychological challenges of spaceflight and lunar surface activities. Their skills and coordination will be critical for mission success.

A Symphony of Spacecraft

Reaching the Moon and investigating its surface needs a complex interplay of properly constructed spacecraft:

Space Launch System (SLS): The most powerful rocket ever built, SLS will deliver the tremendous thrust needed to push the Orion spacecraft and its crew out of Earth's gravity well.

Orion spaceship: This state-of-the-art spaceship serves as the crew's home away from home throughout their journey to and from the Moon. It provides a safe habitat with enhanced life support systems, navigation capabilities, and a docking port for rendezvous with the lunar lander.

Starship Human Landing System (HLS): Developed by SpaceX, Starship HLS is a revolutionary reusable lander responsible for bringing the chosen astronauts from lunar orbit to the surface and back. This futuristic ship will illustrate the possibilities for reusable spacecraft in future lunar expeditions.

A Week on the Lunar Surface

Following a multi-day voyage to lunar orbit, two astronauts will transfer to the Starship HLS for the drop to the South Pole. Once reaching the lunar surface, they will continue on a week-long excursion, undertaking a number of scientific studies. Equipped with next-generation spacesuits, they will execute up to four moonwalks, going out into the alien landscape.

Their activities will include:

Collecting lunar samples: These samples will be studied to understand the Moon's geological history and makeup.
Deploying scientific instruments: These instruments will gather data on the lunar environment, such as seismic activity, the presence of water ice, and the radiation levels.
Testing new technologies: This may involve experimenting with novel strategies for resource utilization or showcasing the capabilities of improved lunar rovers.

Beyond the Landing: A Legacy for the Future

Artemis III is much more than a landing mission. It's a stepping stone for future lunar exploration and a key test bed for technologies needed for establishing a permanent human presence on the Moon. The expertise acquired from this mission will inform future lunar outposts, paving the road for the creation of a viable lunar base. Additionally, Artemis III serves as a springboard for future trips to Mars, allowing us to hone the abilities and technology necessary for deeper space exploration.

This is just the beginning of the Artemis III narrative. This book will dig deeper into the facts of the mission, the challenges and achievements of the crew, and the groundbreaking scientific findings it seeks to achieve. With Artemis III, humanity is prepared to rewrite the history of lunar exploration, venturing further than ever before and claiming a permanent presence on our celestial neighbor.

Mission Objectives and Timeline

A Journey of Discovery and Innovation

Artemis III, planned for launch no sooner than September 2026, is a vital phase in humanity's return to the Moon. This ambitious mission intends to achieve a multiplicity of objectives, establishing the groundwork for a prolonged lunar presence and pushing the boundaries of scientific discovery.

Primary Objectives:

Lunar Landing: First Woman and Person of Color: The mission's most symbolic goal is to be the first to land a woman and a person of color on the Moon's surface. This historic achievement highlights a dedication to diversity and inclusiveness in space exploration, motivating future generations.

Scientific Investigations: Artemis III astronauts will undertake a number of scientific experiments in the Moon's South Pole. Studying this region, rich in potential water ice deposits, carries tremendous significance.

Water Ice Exploration: The major scientific purpose is to explore the presence, distribution, and composition of water ice deposits in permanently shadowed craters. This ice could be a vital resource for future lunar residents, offering drinking water, fuel generating capabilities, and radiation shielding.

Lunar Geology and Environment: Collecting lunar samples and installing scientific instruments will give light on the Moon's geological history, composition, and contemporary environment. Understanding earthquake activity, radiation levels, and surface features will be vital for future lunar communities.

Technology Demonstration: Artemis III serves as a platform to test and validate innovative technologies vital for future lunar exploration. This could include:

Advanced Spacesuits: The newest generation of spacesuits will be put to the test, guaranteeing its functionality and safety for astronauts doing moonwalks.

Lunar Resource Utilization: Technologies for obtaining and exploiting lunar resources, such as water ice, might be shown to pave the way for self-sustaining lunar bases.

Autonomous and Remotely Operated Systems: Testing prototypes of rovers and other robotic systems developed for lunar exploration will play a significant role in future missions.

Mission Timeline (estimated):

Pre-Launch (1-2 years prior):

Astronaut selection and rigorous training regimen covering physical conditioning, spacecraft operation, and lunar surface tasks.

Finalization of spacecraft development and thorough testing of all systems - SLS, Orion, and Starship HLS.

Refining the scientific experiments and instrument deployment plan based on latest research and mission objectives.

Launch and Lunar Transit (7-10 days):

Launch of the Orion spacecraft atop the SLS rocket from Kennedy Space Center.
Multi-day voyage to lunar orbit with crew monitoring systems and making maneuvers.

Lunar Orbit Operations (6.5 days):

Orion docks with the pre-positioned Starship HLS in lunar orbit.
Two astronauts transfer to the Starship HLS while the remaining crew members continue aboard Orion.
The uncrewed Orion spacecraft remains in a Near-Rectilinear Halo Orbit (NRHO) around the Moon for the duration of the landing and surface investigation.

Lunar Landing and Surface Exploration (7 days):

Starship HLS descends to the pre-designated landing spot in the South Polar Region.
Two astronauts complete up to four moonwalks totaling approximately 28 hours.
Activities include: Collecting lunar samples at various sites.
Deploying scientific instruments for data collection.
Testing new technology and conducting lunar surface investigations.
Operating a pre-positioned rover for expanded reach and sample collecting.

Lunar Ascent and Return Transit (4-5 days):

Starship HLS pulls off from the lunar surface, rendezvousing and docking with Orion in lunar orbit.
The team transfers back to Orion for the return flight to Earth.

Orion navigates back towards Earth with crew executing final maneuvers and prepares for re-entry.

Earth Re-entry and Recovery (many hours):

Orion re-enters Earth's atmosphere at great speed, utilizing its heat shield for protection.
The spacecraft splashes down in the Pacific Ocean, where rescue personnel will be waiting to aid the astronauts.

Post-Mission Activities:

Extensive medical exams and debriefing sessions for the astronauts.
Analysis of obtained lunar samples and data gathered from deployed devices.
Evaluation of tested technologies and their possible applicability in future missions.

A Stepping Stone for the Future

The successful completion of Artemis III will constitute a momentous achievement for human space exploration. It paves the way for a permanent human presence on the Moon, showing the possibility of long-term lunar exploration missions. The knowledge and experience gathered from this mission will be crucial for future undertakings, including the development of a lunar outpost and potentially crewed expeditions to Mars.

Evolution of Artemis Program

The Artemis program, NASA's ambitious ambition to return humans to the Moon, hasn't always been the streamlined mission we see today. Its goals and timelines have experienced many revisions since its launch in 2017. Let's look into the intriguing evolution of Artemis, a monument to human inventiveness and the ever-changing terrain of space travel.

From EM-3 to Artemis: A Change in Course (2017-2019)

The story begins in December 2017 with the Space Policy Directive 1, signed by the Trump administration. This directive envisioned a crewed lunar campaign spearheaded by the Orion Multi-Purpose Crew Vehicle (MPCV) and a space station orbiting the Moon named the Lunar Gateway. Initially named Exploration Mission-3 (EM-3), the mission aimed to send four astronauts into a lunar orbit and deliver two key components to the Gateway: the ESPRIT (European System Promoting the Robotic Exploration of Time) deep-space habitat and the U.S. Utilization Module (USM) for crew living quarters.

However, by May 2019, plans had evolved dramatically. ESPRIT and USM were intended to travel separately on a commercial launch vehicle, streamlining EM-3. This updated project, now designated Artemis 3, planned for an expedited timeframe, seeking a lunar landing by the end of 2024. Envisioned was a rendezvous between the Orion MPCV and a basic Gateway with only the Power and Propulsion Element (PPE) and a small habitat/docking node. This Gateway would be deployed after the Artemis 3 landing.

Shifting Strategies and Landing Site Selection (2019-2020)

By early 2020, the plan received another change. The concept of Orion and the Human Landing System (HLS) rendezvousing with the Gateway was abandoned in favor of a more direct approach - Orion docking directly with the HLS. The Gateway's delivery was thus pushed to after Artemis 3. This streamlined method aims to accelerate the first crewed lunar landing.

The landing site also became a matter of focus. Initially, Artemis missions weren't bound to a specific lunar site. However, in 2019, the South Polar Region emerged as the chosen landing site for Artemis 3. The rationale was compelling: the presence of possible water ice accumulation in permanently shaded craters. This ice could be a vital resource for future lunar communities, offering drinking water, fuel production capabilities, and radiation shielding.

Technical Hurdles and Delays (2020-Present)

As the development developed, technical obstacles came to light. Delays emerged owing to difficulties with spacesuit development and the Starship HLS landing system under developed by SpaceX. An Office of Inspector General assessment in August 2021 revealed that the spacesuits wouldn't be available until April 2025 at the earliest, putting back the intended late 2024 launch date for Artemis 3.

The ensuing months witnessed further modifications. By November 2021, NASA Administrator Bill Nelson confirmed a launch no early than 2025. In June 2023, Jim Free, NASA's assistant administrator for exploration systems development, suggested a launch "probably" no early than 2026. Finally, in January 2024, NASA formally delayed Artemis 3 to no early than September 2026.

However, amidst these delays, there have been bright spots. Starship's successful orbital trip in March 2024 highlighted substantial advancement. Additionally, NASA disclosed the scientific equipment chosen for Artemis III, demonstrating the mission's commitment to scientific discovery.

A Program in Progress: The Road Ahead

The Artemis program continues to expand, adjusting to technological hurdles and shifting goals. Despite the delays, the program's basic goals remain: creating a sustained human presence on the Moon and exploiting it as a stepping stone for future exploration activities, including Mars missions. As the Artemis program develops, we should expect incremental refinements, technological advancements, and perhaps even new discoveries that may lead to further expansion of the mission objectives.

We will dig further into the current condition of the Artemis program, analyzing the planned missions, the obstacles and opportunities that lie ahead, and the great potential this program offers for the future of space exploration. Join us as we witness the culmination of years of effort and witness the birth of a new era on the Moon.

Chapter 2

Mission Planning and Development

Artemis III isn't only a space mission; it's a perfectly orchestrated ballet of technology, human brilliance, and precise planning. This chapter goes into the intricate realm of mission planning and development, studying the obstacles and successes involved in bringing Artemis III to life.

Defining Objectives and Setting the Stage

The journey begins with a clear vision. NASA, along with its international partners, outlined the key objectives for Artemis III: landing the first woman and person of color on the Moon, conducting essential scientific research at the lunar South Pole, and testing technology crucial for future lunar exploration.

Selecting the Landing Site: A Quest for Water Ice

Choosing the landing place was a vital decision. After thorough investigation, the South Polar Region emerged as the leader. This location boasts perpetually shadowed craters, where temperatures plunge to -243°C (-405°F). These freezing circumstances could contain a treasure trove — water ice deposits. The presence of water ice would be a game-changer, providing a key resource for future lunar residents. It might be utilized for drinking water, creating rocket fuel through electrolysis, and even as radiation shielding for lunar settlements.

Crafting the Crew: A Team for the Ages

Selecting the astronauts who will make the historic journey to the Moon is a tough process. Candidates undertake thorough medical and psychological examinations, physical conditioning regimens, and intensive training in spacecraft systems, lunar surface operations, and scientific processes. The chosen crew will represent a varied group, a testament to the expanding panorama of space research.

Building the Machines: A Trio of Spacecraft

No single spacecraft can meet the multiple goals of Artemis III. Instead, a complicated interplay of three separate vehicles is required:

Space Launch System (SLS): The most powerful rocket ever built, SLS acts as the muscle, delivering the immense thrust needed to launch the Orion spacecraft and its crew out of Earth's gravity well.

Orion spaceship: This state-of-the-art spaceship serves as the crew's home away from home during their journey. It provides a safe habitat with enhanced life support systems, navigation capabilities, and a docking port for rendezvous with the lunar lander.

Starship Human Landing System (HLS): Developed by SpaceX, Starship HLS is the revolutionary, reusable lander responsible for bringing the chosen astronauts from lunar orbit to the surface and back. Its reusability provides the possibility for a more sustainable lunar exploration strategy.

Developing this spacecraft requires years of painstaking engineering, intensive testing, and international collaboration. Every system, every component, must perform flawlessly for mission success.

Harnessing Science: Instruments for Discovery

Scientific exploration is a cornerstone of Artemis III. Several cutting-edge devices will be installed on the lunar surface to acquire invaluable data:

Lunar Environment Monitoring Station (LEMS): This instrument will probe deep into the mysteries buried beneath the lunar surface, revealing insights on the Moon's internal structure and composition.

Lunar Effects on Agricultural Flora (LEAF): This experiment will study the possibilities for cultivating crops in the harsh lunar environment. Understanding the influence of lunar gravity and radiation on plant life is crucial for growing self-sustaining lunar bases.

Lunar Dielectric Analyzer (LDA): This device will investigate the electrical properties of lunar regolith, the loose soil covering the Moon's surface. This knowledge can offer light on the Moon's geological history and mineral possibilities.

These devices are the culmination of years of study and development. Their successful deployment and data collecting will add greatly to our understanding of the Moon.

Training for Success: Simulating the Lunar Experience

Preparing the astronauts for the physical and psychological demands of spaceflight is crucial. They train in specially constructed simulators, imitating the Orion ship environment and the lunar surface activities. Additionally, they participate in geological field investigations on Earth in regions like volcanic terrains, replicating the lunar landscape. These immersive training sessions ensure the astronauts are able to handle unexpected scenarios and make vital decisions throughout the journey.

Planning for Every Contingency: Risk Management

Spaceflight is fundamentally dangerous. The Artemis III mission preparation team doesn't shy away from these hazards; they embrace them. Identifying possible risks and devising mitigation methods is key. This encompasses eventualities including equipment faults, medical issues during the voyage, and unanticipated lunar surface conditions. Through thorough planning and contingency measures, the team aims to assure mission success and the safety of the crew.

International Collaboration: A Global Endeavor

Artemis III is far more than a lone nation's attempt. It's a monument to the strength of worldwide collaboration. Partners like the European Space Agency (ESA) and the Canadian Space Agency (CSA) are contributing important expertise and technologies to the mission. ESA's Orion Service Module offers propulsion capabilities for the Orion spacecraft, while CSA's Canadarm3 robotic arm will be important in future lunar development
Artemis 3 Mission Plan

The Price of Progress: Overcoming Technical Challenges

No large-scale space program is without its problems. Artemis III confronts its own set of obstacles, particularly in the domain of technology development.

Spacesuit creation: One recurring difficulty has been the creation of next-generation spacesuits for lunar surface exploration. These suits need to be lightweight, flexible enough for a wide range of mobility, and provide enough protection from the harsh lunar environment - severe temperatures, radiation, and micrometeoroids. Delays in spacesuit development have led to the overall mission timeline modifications.

Starship HLS Development: Starship HLS, the reusable lunar lander, marks a technological leap forward. However, its development has needed complicated engineering and considerable testing. While the successful March 2024 orbital flight was a welcome milestone, further testing and refining are needed to guarantee its safe operation for a crewed lunar landing.

Adapting and Overcoming: The Evolving Timeline

These technical hurdles have prompted revisions to the intended Artemis III timeframe. Initially aiming for a landing by the end of 2024, the mission has been pushed back to no early than September 2026. While delays are never desirable, they allow for extensive testing and ensure the safety of the crew and mission success.

The Road Ahead: A Sustainable Future on the Moon

Despite the hurdles, Artemis III sits on the cusp of a monumental feat. This mission lays the groundwork for a more sustained human presence on the Moon. Here's what lies ahead:

creating a Lunar Basecamp: The knowledge and experience gathered from Artemis III will be important for creating a long-term lunar outpost. This installation might function as a scientific research outpost, a refueling station for deep space exploration, and a stepping stone for future journeys to Mars.

Resource Utilization: Technologies to extract and utilize lunar resources like water ice will be further developed. This could transform the way future lunar colonies operate, boosting self-sufficiency and lowering dependence on Earth-based resources.

Technological Advancements: Artemis III will serve as a testbed for new lunar exploration technology. Lessons obtained will inform the development of more advanced rovers, autonomous exploration systems, and in-situ resource utilization (ISRU) strategies.

A Journey of Discovery and Inspiration

Artemis III is considerably more than just a lunar landing mission. It's a combination of human creativity, scientific curiosity, and the persistent urge to explore. This book will go deeper into the intricate intricacies of mission development, the hurdles conquered, and the technology that will catapult humanity further into the universe. We will meet the astronauts who will make this historic mission and study the scientific findings they expect to achieve. Join us as we witness the birth of a new age in human space exploration and take a great leap towards a lunar future.

Crew Selection and Training

The Artemis III mission rests on the remarkable abilities and unshakable courage of its crew. Sending astronauts to the Moon isn't just about operating a spaceship; it's about doing cutting-edge science in a tough environment. This chapter discusses the laborious process of selecting and training the Artemis III astronauts, the pioneers who would rewrite history on the lunar surface.

Building a Diverse Team: Breaking Barriers in Space

Artemis III is committed to diversity, seeking to be the first mission to land the first woman and the first person of color on the Moon. This momentous feat transcends symbolism; it encourages future generations and dismantles barriers in the historically homogenous field of spaceflight.

Selecting the crew begins with a global pool of qualified astronauts. Candidates from NASA, its international partners like the European Space Agency (ESA) and the Canadian Space Agency (CSA), and maybe even the developing private space sector will be considered. The chosen crew will presumably consist of four highly skilled individuals:

Mission Commander: This astronaut acts as the leader of the team, responsible for overall mission success. They possess considerable aviation experience, great leadership qualities, and the ability to make vital judgments under pressure.

Pilot: The pilot carefully manages the Orion spacecraft during numerous mission phases, from launch to docking and re-entry. They possess in-depth understanding of spacecraft technologies and a cool disposition for addressing difficult situations.

Mission Specialist 1 (MS-1): This astronaut serves as the "engineer" of the crew, responsible for monitoring spacecraft systems and completing technical tasks. They have a deep understanding of spacecraft engineering and a thorough approach to problem-solving.

Mission Specialist 2 (MS-2): This astronaut acts as the "geologist" of the crew, leading science activities during the lunar surface expedition. They possess a foundation in geology or a related field and skill in running scientific instruments.

A Rigorous Selection Process:

Becoming an Artemis III astronaut is no simple accomplishment. Candidates undertake a thorough selection process involving numerous critical elements:

Medical Evaluation: Extensive medical exams assure the candidate's physical and psychological readiness for the challenging spaceflight environment.

Psychological Evaluation: Astronauts undergo great isolation, high-pressure circumstances, and long durations of confinement. Evaluations examine mental resilience, teamwork abilities, and the ability to operate under stress.

Technical Expertise: Candidates must possess strong academic backgrounds in science, engineering, or mathematics, combined with flight experience and technical skills appropriate to their given tasks.

Interview Process: A series of rigorous interviews assess leadership qualities, problem-solving talents, communication skills, and the ability to work effectively as part of a team.

Training for Success: Simulating the Lunar Experience

Once selected, the Artemis III astronauts engage on a multi-year, intensive training regimen. This program encompasses:

Physical Conditioning: Astronauts endure intensive physical conditioning to maintain optimum health and fitness in the low-gravity environment of space. Training involves weightlifting, cardiovascular training, and simulated spacewalks with specialized suits.

Systems Training: In-depth training familiarizes the crew with the intricacies of the Orion spacecraft, Starship HLS lander, and all the research instruments they will utilize. This includes practicing launch procedures, emergency protocols, and docking maneuvers in simulators.

Geological Field Studies: Training missions to volcanic terrains, deserts, and other severe conditions resemble the lunar surface.

Astronauts rehearse moonwalk techniques, geological sample collection procedures, and deploying scientific instruments under field settings.

collaboration and Communication Training: Crew members participate in intense collaboration and communication training to develop a cohesive unit. This includes practicing collaboration under duress and improving skills for successful communication with mission control on Earth.

Living in Isolation: Psychological Preparation

The psychological toll of spaceflight shouldn't be underestimated. The Artemis III crew will undertake rigorous training to cope with the mental problems of isolation and confinement during the long lunar mission. Techniques include mindfulness exercises, stress management measures, and keeping communication with loved ones back on Earth.

A World Watching: The Weight of Expectation

The Artemis III crew not only takes the technical and scientific load of the mission, but also wears the weight of global public attention. They will be ambassadors for humanity, inspiring future generations with their courage and dedication. Understanding the significance of their work, the crew will participate in media training and public outreach initiatives.

A Team Forged in Fire: Ready for the Challenge

The precisely picked and rigorously trained Artemis III crew marks the result of their efforts. They are a team of remarkable individuals, united by a single aim – to push the boundaries of human exploration and write their names in the history books. As they embark on their trip to the Moon, they carry the hopes and dreams of humanity, setting the path for a future where lunar exploration becomes normal. While the technical components of crew selection and training are critical, it's important to consider the human factor. Let's meet some of the probable crew members who might be selected for this historic mission:

Dr. Amelia Sharma (Mission Specialist 2): A bright geologist with a passion for lunar exploration, Dr. Sharma has dedicated her career to investigating meteorites and volcanic rocks for clues about the Moon's genesis. Her expertise in lunar geology makes her a great contender for the MS-2 post, responsible for leading research activities on the lunar surface. Her calm manner and precise approach ensure she can properly gather samples and deploy instruments in the challenging lunar environment.

Captain Kai Tanaka (Pilot): A seasoned pilot with nerves of steel, Captain Tanaka has flown multiple missions aboard the International Space Station (ISS). His outstanding piloting skills and unshakable focus are essential tools for the task. Calm under pressure, Captain Tanaka excels at handling complicated situations and guarantees the safe navigation of the Orion spacecraft during critical mission phases like launch, docking, and re-entry.

Lieutenant Colonel Nadia Petrova (Mission Specialist 1): A highly talented engineer with a background in robotics, Lieutenant Colonel Petrova adds a lot of experience to the crew. Her knowledge in spacecraft systems and her ability to handle technical problems make her a perfect candidate for the MS-1 post. During the mission, she will be responsible for monitoring the health of the Orion spacecraft and Starship HLS lander, ensuring all systems perform efficiently for a successful mission.

Dr. David Hernandez (Mission Commander): A senior astronaut with a stellar career, Dr. Hernandez has vast experience in spacewalks and scientific research onboard the ISS. His outstanding leadership qualities, paired with his technical expertise and calm manner, make him a perfect option for the Mission Commander post. Dr. Hernandez will be accountable for the overall success of the mission, making important decisions and ensuring the crew performs cohesively as a unit.

These are only a few instances, and the actual crew selection process will likely entail a far broader pool of competent people. However, they provide a glimpse into the different skillsets and backgrounds that will be represented on Artemis III.

Beyond Training: Building Cohesion

Training isn't only about technical proficiency; it's also about developing a solid team chemistry. The crew will undergo extensive collaboration activities to foster trust and communication. Sharing meals together, partaking in social activities, and completing psychological therapy sessions all contribute to building a cohesive unit that can effectively face the obstacles and demands of the mission.

The Pressure of the Spotlight: Public Engagement

The Artemis III crew will be catapulted into the global limelight. They will participate in media training to polish their communication skills and communicate with the public in a way that inspires future generations. Sharing their experiences, the science underpinning the mission, and their personal motivations will captivate the public's imagination and generate enthusiasm for space exploration.

A Legacy of Inspiration: Leaving Their Mark

The Artemis III crew isn't just a squad of astronauts; they're pioneers. They will be the first people to walk on the Moon in nearly half a century, leaving their footprints and their memories inscribed in lunar history. Their voyage will inspire many individuals throughout the world, illustrating the strength of human aspiration and the unlimited possibilities of space exploration.

As launch day approaches, the world will be watching with bated breath, cheering on these brave individuals as they embark on a historic journey that promises to rewrite the story of human space travel.

Spacecraft Overview: Orion and Starship HLS

The Artemis III mission will rely on two crucial spacecraft to execute its historic lunar landing: the Orion capsule and the Starship Human Landing System (HLS).

1. Orion Spacecraft: The Crew's Home Away from Home

[Image of Orion Spacecraft (Artemis)]

The Orion spacecraft serves as the crew's comfortable and secure abode throughout their journey to lunar orbit and back to Earth.

Developed by a consortium lead by Lockheed Martin, Orion is a next-generation capsule designed for deep space exploration missions.

Here's a rundown of Orion's primary features:

Crew Capacity: Four astronauts
Length: 32.7 feet (10 meters)
Diameter: 16.5 feet (5 meters)
Habitable Volume: 316 cubic feet (8.95 cubic meters) - approximately 1.5 times larger than the Apollo capsules

Orion consists of two primary modules:

Crew Module (CM): This pressurized module provides living accommodations for the crew during the journey. It has advanced life support systems, pleasant sitting arrangements, a mini-kitchen, and a hygiene station. Large windows offer panoramic views of Earth and space.

Service Module (SM): This unpressurized module houses the main propulsion system, generates electrical power through solar arrays, and stores water and other supplies needed for the mission. The European Space Agency (ESA) contributes the Orion Service Module.

2. Starship Human Landing System (HLS): From Lunar Orbit to the Surface and Back

[Image of Starship Human Landing System (Artemis)]

Starship HLS, designed by SpaceX, is a revolutionary reusable lander responsible for ferrying humans between lunar orbit and the surface. This single-stage lander uses the power of SpaceX's Starship launch vehicle.

Here's a closer look at Starship HLS:

Crew Capacity: Up to 4 astronauts
Height: 169 feet (51.5 meters)
Diameter: 30 feet (9 meters)

Starship HLS is a version of the bigger Starship launch vehicle. Key aspects include:

Raptor Engines: Six Raptor engines installed on the lander provide the thrust needed for descent to the lunar surface and ascension back to lunar orbit for rendezvous with Orion.

Landing Gear: Deployable landing legs will cushion the Starship HLS upon touchdown on the lunar surface.

Docking Port: A docking port on the top of the lander aids crew transfer between Starship HLS and Orion.

Cargo Bay: The Starship HLS features a cargo bay that can transfer research equipment, lunar samples, and other supplies to and from the lunar surface.

A Powerful Duo for Lunar Exploration

Orion and Starship HLS offer a powerful collaboration for crewed lunar missions. Orion provides a safe and comfortable environment for the crew throughout the long trip, while Starship HLS offers a versatile and reusable landing system for lunar surface exploration. Together, these spacecraft lay the way for a sustainable human presence on the Moon.

Chapter 3

Launch and Transit

The Artemis III mission will climax in a dramatic launch, sending humans on a multi-day trek to lunar orbit. We dig into the minute intricacies of the launch sequence, the challenges of space flight, and the crew's activities during their passage to the Moon.

Liftoff: A Powerful Ascent

The show begins at the Kennedy Space Center (KSC) in Florida. The Orion spacecraft, nestled atop the powerful Space Launch System (SLS) rocket, awaits its fiery baptism. The SLS, the most powerful rocket ever built, delivers a tremendous blast as it ignites its four RS-25 engines and twin solid rocket boosters. The launchpad trembles with the raw strength, and the vehicle ascends into the sky, leaving a path of fire and smoke in its wake.

Separation and Ascendancy: Orion Takes Flight

After reaching a specified height, the solid rocket boosters, having consumed their fuel, separate from the core stage of the SLS in a stunning display. Shortly later, the core stage itself separates, its duty accomplished. Orion, now propelled by its own engine in the service module, begins its trip towards the Moon.

Translunar Injection: Setting Course for Lunar Orbit

Following Earth orbit insertion, Orion undertakes a vital maneuver called the Translunar Injection (TLI). During TLI, the spacecraft's engine operates for a specific period, adjusting its velocity and sending it on a course towards the Moon. This burn establishes the spacecraft's lunar transfer orbit, a path that will carry it around the Earth and eventually towards the Moon's gravitational influence.

Settling In: Life Aboard Orion

Once on the lunar transfer trajectory, the crew switches from the launch phase to the long-duration spaceflight routine. They adapt to the microgravity environment, a state of weightlessness that involves adaptations to movement and daily activities.

The crew's day is methodically scheduled, mixing critical work with scientific research and personal cleanliness. Activities include:

Monitoring Systems: Astronauts regularly monitor spaceship systems, ensuring everything performs effectively throughout the journey.
Conducting Experiments: The crew might participate in pre-planned scientific experiments or educational outreach activities sent back to Earth.

Exercising: Regular physical exercise is necessary to preserve muscle tone and bone health in microgravity. The Orion ship contains a dedicated exercise facility for astronauts to occupy.

Meals and Hygiene: Astronauts enjoy specially made meals especially for space travel. The spacecraft also features specialized areas for personal hygiene and garbage management.

Navigating the Journey: Course Corrections and Mid-Course Maneuvers

Throughout the transit, the crew and mission control work together to monitor the spacecraft's course. Minor adjustments in course, called mid-course maneuvers (MCMs), may be essential to achieve an exact lunar orbit insertion. These MCMs are often executed utilizing the Orion's service module engine.

A Glimpse of Home: Earth Observations and Lunar Rendezvous

As the spaceship advances on its lunar trajectory, astronauts get a chance to observe Earth shrink into a stunning blue marble against the backdrop of the cosmos. They might also capture magnificent photographs of the lunar surface as it gradually grows larger in their vision.

Lunar Orbit Insertion (LOI): Reaching the Lunar Neighborhood

Nearing the Moon, Orion undertakes another key maneuver — the Lunar Orbit Insertion (LOI). The service module engine ignites again, adjusting the spacecraft's velocity and setting it into a stable orbit around the Moon. This lunar orbit will serve as the staging site for the future lunar landing.

Docking with Starship HLS: Preparation for Descent

While in lunar orbit, Orion awaits the arrival of the Starship HLS. This uncrewed Starship HLS, launched separately from Earth, will rendezvous and dock with Orion utilizing its autonomous docking technology. Once docked, the crew will transfer from Orion to the Starship HLS in preparation for their historic lunar descent.

A Turning Point: The Journey from Earth to Lunar Orbit

The launch and transit phase of Artemis III is a key and awe-inspiring chapter in the mission. It is the result of years of planning, meticulous engineering, and rigorous crew training. As Orion enters lunar orbit, humanity stands on the edge of a tremendous achievement - the first crewed lunar landing in over five decades. The stage is now set for the descent to the lunar surface and the unprecedented scientific investigation that awaits the Artemis III astronauts.

Space Launch System (SLS) Overview

The Space Launch System (SLS), often known as the SLS rocket, is an American super heavy-lift expendable launch vehicle designed by NASA. It is the most powerful rocket ever developed and is meant to launch personnel and cargo on deep space missions, including Artemis trips to the Moon and future missions to Mars.

The SLS is a sophisticated system that comprises of several major components:

Core Stage: The core stage is the heart of the SLS rocket and generates the most thrust. It is propelled by four RS-25 engines that were originally built for the Space Shuttle program. The core stage also holds the vehicle's avionics, guidance systems, and propellant tanks.

Solid Rocket Boosters (SRBs): Two SRBs are positioned on either side of the core stage and provide additional thrust during launch. The SRBs are the largest ever produced for a space launch vehicle and are jettisoned a few minutes after liftoff.

Interim Cryogenic Propulsion Stage (ICPS): The ICPS is the upper stage of the SLS rocket and is powered by a single RL10 engine. The ICPS is responsible for injecting the Orion spacecraft or other payloads into their appropriate trajectory after separation from the core stage.

Exploration Upper Stage (EUS): The EUS is a more powerful upper stage that is currently under development. It will eventually replace the ICPS and will be capable of producing additional push for flights to deep space destinations.

Orion Spacecraft: The Orion spacecraft is not officially part of the SLS, but it is the crew capsule that will be launched on top of the SLS for Artemis missions. Orion will provide a safe habitat for astronauts during their travel to deep space and their return to Earth.

The SLS is a gigantic rocket that reaches 322 feet (98 meters) tall and has a diameter of 27.6 feet (8.4 meters). It is capable of delivering a whopping 8.8 million pounds of thrust at liftoff, which is more thrust than any other operational rocket in the world.

The maiden launch of the SLS took place on November 16, 2022, as part of the Artemis 1 mission. The Artemis 1 mission was an uncrewed test flight that successfully sent the Orion spacecraft on a loop around the Moon and back to Earth. The next launch of the SLS is planned for the Artemis 2 mission in 2024, which will be the first crewed voyage around the Moon since Apollo 17 in 1972.

The SLS is a crucial component of NASA's Artemis program and is required for future missions to deep space. It is a strong and adaptable rocket that will serve to bring in a new era of space exploration.

Launch Site and Vehicle

The Artemis III mission rests on two essential elements: the launch site and the launch vehicle. This chapter digs into the complex intricacies of Kennedy Space Center (KSC) - the historic launchpad from which Artemis III will take flight – and the gigantic Space Launch System (SLS) rocket, the colossal machine tasked with propelling men towards the Moon.

Kennedy Space Center (KSC): A Legacy of Exploration

Nestled on Merritt Island, Florida, Kennedy Space Center (KSC) is a hallowed site in the history of space exploration. It served as the launching location for landmark missions like Apollo 11, the first crewed lunar landing, and continues to be a crucial site for NASA's activities.

KSC possesses the required infrastructure to support the launch of the SLS, including:

Launch Complex 39B (LC-39B): This legendary launch pad, expertly rebuilt for the SLS era, will serve as the departure site for Artemis III. LC-39B contains a flame trench to deflect the extreme heat created during launch, a mobile launcher umbilical tower that offers vital connections and fueling capabilities for the SLS, and a crew access arm that allows astronauts to securely join Orion prior to liftoff.

Vehicle Assembly Building (VAB): This huge structure serves as the assembly point for the SLS rocket. Here, the various stages of the SLS, together with the Orion spacecraft, are methodically built, integrated, and tested before being transported to the launch pad.

Launch Control Center (LCC): This nerve center houses the mission control crew responsible for managing the launch countdown, monitoring spacecraft telemetry during ascent, and guaranteeing the mission's smooth execution.

The Space Launch System (SLS): A Colossal Force for Deep Space
The SLS is the throbbing heart of the Artemis III launch. This super heavy-lift expendable launch vehicle, the most powerful ever developed, contains the massive thrust necessary to drive the Orion spacecraft and its crew on their lunar voyage. Here's a rundown of the essential components of this engineering marvel:

Core Stage: This central stage holds the four powerful RS-25 engines, originally developed for the Space Shuttle program. These engines use liquid hydrogen and liquid oxygen propellants to create an incredible 8.8 million pounds of force at liftoff. The core stage also holds the vehicle's avionics, guidance systems, and propellant tanks.

Solid Rocket Boosters (SRBs): Two SRBs flank the core stage, delivering an extra thrust surge during the early ascent phase. These boosters, the largest ever developed for a space launch vehicle, burn solid propellant and are ejected a few minutes after liftoff.

Interim Cryogenic Propulsion Stage (ICPS): This upper stage ignites following the core stage separation and SRB jettison. Powered by a single RL10 engine, the ICPS propels the Orion spacecraft out of Earth's orbit and sends it on a route towards the Moon.

A Symbiotic Relationship: Launch Site and Launch Vehicle

The success of the Artemis III launch rests on the smooth connection between KSC and the SLS. KSC offers the required infrastructure, the launch pad, and the knowledge to precisely prepare and launch the mighty SLS. The SLS, in turn, generates the tremendous energy required to overcome Earth's gravity and carry the Artemis III crew towards their lunar objective.

A Legacy in the Making:

The launch of Artemis III from KSC aboard the SLS will mark a landmark occasion. It reflects humanity's continuous pursuit of lunar exploration and lays the path for a more sustainable human presence on the Moon. The launch location and the launch vehicle, together, constitute the motors that fuel humanity's goals of deep space travel.

Journey to Lunar Orbit

The Artemis III mission doesn't begin with liftoff; it's a multi-day odyssey that starts with painstaking pre-launch preparations and concludes in a magnificent arrival into lunar orbit. This chapter digs into the intricate details of this vital period, detailing the crew's activities, the hardships of space flight, and the awe-inspiring vistas witnessed along the route.

T-Minus Several Days: Launch Preparations

Long before liftoff, a frenzy of activity takes place at Kennedy Space Center (KSC). The Orion spacecraft sits atop the powerful Space Launch System (SLS) undergoes rigorous final testing. The crew arrives at KSC, partaking in briefings, medical examinations, and mental exercises to ensure they are physically and mentally prepared for the arduous journey ahead.

Liftoff: A Powerful Ascent

Launch day dawns, and anticipation hangs thick in the air. Millions around the world tune in to witness the spectacle. Astronauts board Orion through the crew access arm, fastened securely into their seats. With a loud scream and a blazing smoke, the SLS ignites its four RS-25 engines and twin solid rocket boosters. The launchpad trembles with the raw strength as the vehicle ascends, leaving a path of fire and smoke in its wake.

Max-Q: The Moment of Peak Pressure

A few minutes into the trip, the vehicle reaches a critical point termed Max-Q. This is the moment when the spacecraft experiences the maximum dynamic pressure of the whole ascent. The crew can sense a modest g-force increase as the vehicle pushes through the thickest region of the atmosphere.

Separation Drama: Core Stage and Solid Rocket Boosters Detach

After reaching a specified altitude, the drama unfolds. The solid rocket boosters, having consumed their propellant, separate from the core stage with a series of explosive charges. The expended rockets parachute back to Earth for retrieval and probable reuse in future missions.

Core Stage Separation and Orion Takes Over

Shortly after the SRB separation, the core stage, having fulfilled its function, likewise departs from the Orion spacecraft. The Orion's service module engine starts, moving the spaceship away of the spent core stage's path.

Translunar Injection (TLI): Setting Course for the Moon

Once in Earth orbit, Orion undertakes a vital maneuver called the Translunar Injection (TLI). The service module engine fires for a predetermined period, adjusting the spacecraft's velocity and placing it on a trajectory towards the Moon. This drives Orion out of Earth's orbit and sets the spacecraft's lunar transfer orbit, a path that will carry it around the Earth and eventually towards the Moon's gravitational influence.

Adapting to Microgravity: Life Aboard Orion

As Orion exits Earth's hold, the crew moves from the high-g forces of launch to the microgravity environment of spaceflight. This feeling of weightlessness needs modifications to movement and daily routines. Astronauts spend the opening hours acclimating to microgravity, having a "floaty" sensation and needing to relearn routine actions like eating and sleeping.

Settling into the Routine: A Well-Oiled Machine

The crew's day becomes a well-rehearsed routine, mixing critical responsibilities with scientific research and personal cleanliness. Activities include:

Monitoring Systems: Astronauts closely monitor spaceship systems, ensuring all functions optimally throughout the flight.
Conducting Experiments: The crew might participate in pre-planned scientific experiments or educational outreach activities sent back to Earth. These investigations could range from

evaluating new materials in microgravity to viewing cosmic occurrences.

Exercising: Regular physical exercise is necessary to preserve muscle tone and bone health in microgravity. The Orion ship contains a dedicated exercise facility for astronauts to occupy. Resistance bands and carefully specialized training equipment assist prevent the muscle loss associated with microgravity.

Meals and Hygiene: Astronauts enjoy specially made meals especially for space travel. Packaged in single-serve pouches or dried versions requiring rehydration, these meals supply the necessary nourishment for the crew. The spacecraft also features specialized areas for personal hygiene and garbage management.

A Celestial Spectacle: Earthrise and Lunar Views

As Orion advances on its lunar transfer trajectory, the crew sees a stunning view — Earthrise. Our home planet shrinks into a magnificent blue marble against the backdrop of the cosmos, a strong reminder of the vastness of space and the fragility of our globe.

In the ensuing days, the Moon progressively becomes larger in their perspective, morphing from a distant ball into a world with its own unique craters, mountains, and plains. Astronauts might utilize this time to investigate lunar characteristics in preparation for their forthcoming landing.

Mid-Course Maneuvers (MCMs): Fine-Tuning the Trajectory

Throughout the multi-day transit, the crew and mission control work together to monitor the spacecraft's course. Deviations from the specified course may occur due to slight gravitational effects or tiny mistakes during launch. To compensate for these deviations, the crew could undertake mid-course maneuvers (MCMs).

Navigation Techniques: Astronauts rely on a combination of celestial navigation and ground-based tracking data to determine their precise position and course correction needs. Celestial navigation involves utilizing the locations of stars and other celestial bodies as reference points. Ground-based tracking stations on Earth broadcast radio signals that Orion receives and utilizes to establish its whereabouts.

Performing MCMs: If a course adjustment is necessary, the Orion service module engine operates for a brief, calculated period. The direction and length of the burn depend on the magnitude of the trajectory deviation. Multiple MCMs might be done along the journey to achieve a exact lunar orbit insertion.

A Glimpse of the Past: Apollo Era Sites

As Orion flies deeper into space, the crew might have the opportunity to witness landmark spots from the Apollo period. They may fly above craters where previous missions landed, viewing the footprints and hardware left behind by prior lunar pioneers. Gazing at these sites serves as a strong reminder of humanity's past successes in lunar exploration and fuels anticipation for their own momentous landing.

Communications with Earth: Staying Connected

Despite the immense distance, the crew maintains connected to Earth through a sophisticated communication system. They may exchange phone messages, emails, and even live video streams with mission control and their loved ones back home. These transmissions provide a key link to Earth, delivering emotional support and keeping the crew informed of mission progress.

Spacewalks: A Contingency or Opportunity?

While spacewalks aren't a scheduled activity during the normal Artemis III mission, they can't be fully ruled out. In the case of unforeseen circumstances, such as a catastrophic systems malfunction requiring external repair, astronauts might don their spacesuits and execute a contingency spacewalk outside Orion.

Spacesuits: The Artemis program has created next-generation spacesuits specifically designed for lunar exploration. These suits offer greater mobility and flexibility than earlier generations, enabling astronauts with the capacity to conduct complicated activities on the lunar surface during a contingency spacewalk.

It's vital to note that spacewalks during the transit phase are exceedingly complicated and pose significant dangers. They would only be undertaken as a last resort after thorough assessment by mission control and the crew.

A Psychological Journey: Coping with Isolation

The psychological toll of spaceflight shouldn't be underestimated. The crew spends days locked within the relatively small cabin of Orion, distant from the familiarity of Earth. Isolation and confinement can lead to feelings of loneliness, anxiety, and boredom.

Psychological Countermeasures: To mitigate these problems, the crew conducts substantial psychological preparation prior to launch. Relaxation techniques, maintaining connection with loved ones, and following a scheduled daily routine all contribute to preserving mental well-being during the lengthy travel.

Lunar Orbit Insertion (LOI): Reaching the Moon's Grasp

As Orion nears the Moon, it conducts another vital maneuver — the Lunar Orbit Insertion (LOI). The service module engine ignites again, adjusting the spacecraft's velocity and setting it into a stable orbit around the Moon. This lunar orbit will serve when Orion nears the Moon, it undertakes another vital maneuver - the Lunar Orbit Insertion (LOI). The service module engine ignites again, adjusting the spacecraft's velocity and setting it into a stable orbit around the Moon. This lunar orbit will serve as the staging site for the future lunar landing.

The LOI burn can be a breathtaking experience for the crew. Gazing out the viewport, they observe the lunar surface grow gradually larger, exposing its craters, plains, and towering mountains in amazing detail. The successful LOI is a major milestone in the program, signaling humanity's return to lunar orbit after a several-decade break.

Docking with Starship HLS: Preparation for Descent

While in lunar orbit, Orion isn't alone. The Starship Human Landing technology (HLS), launched separately from Earth a few days before, rendezvous and docks with Orion utilizing its autonomous docking technology. This uncrewed Starship HLS carries the lunar lander that will ferry the astronauts down to the surface.

Docking maneuvers are a delicate ballet in space, requiring accurate navigation and close monitoring by both the crew and mission control. Once docked, the crew can finally move from the Orion capsule to the Starship HLS through a specially engineered docking tube.

A Change of Scenery: Inside the Starship HLS

Stepping onboard the Starship HLS is like entering a new spacecraft. Compared to Orion's crew module, the Starship HLS boasts a more roomy and functional cabin. It contains numerous storeys and enormous windows allowing panoramic views of the lunar terrain. The lander also houses crucial supplies and equipment needed for the lunar surface expedition.

A Final Look Back: Earthrise from Lunar Orbit

Before disembarking from Orion and preparing for the lunar descent, the crew might take a minute to enjoy another stunning vista —

Earthrise from lunar orbit. Our home planet looks as a small blue marble suspended against the dark backdrop of space, a stunning reminder of the immensity of the cosmos and the fragility of life. This vista provides as a somber moment of meditation on humanity's role in the universe and the importance of space exploration.

Lunar Descent Preparations: Suit Up and Brief Up

With Orion firmly docked with Starship HLS, the focus goes towards the lunar descent. The crew methodically prepares for their historic walk on the Moon. This preparation involves:

Donning Spacesuits: Astronauts carefully don their advanced extravehicular activity suits (EVAs) designed specifically for lunar exploration. These suits provide life support systems, protection from the hostile lunar environment, and the mobility needed to function effectively on the lunar surface.

Mission Briefing: The crew performs a final mission briefing with mission control, covering the descent plan, lunar surface operations, and emergency procedures. This briefing assures everyone is on the same page and prepared for any possibilities that might emerge during the landing and surface exploration.

A Turning Point: The Gateway to Lunar Exploration

The journey to lunar orbit culminates with the crew prepared and eager to descend to the lunar surface aboard Starship HLS. This event marks a key turning point in the endeavor. The success of the transit phase has created the basis for the historic lunar landing, paving the path for enormous scientific discovery and human exploration on the Moon.

Fine-Tuning the Trajectory

During the multi-day transit from Earth to lunar orbit, the seemingly easy path of the Orion spacecraft requires regular monitoring and occasional changes. This sophisticated process, known as trajectory fine-tuning, guarantees the spacecraft arrives at the Moon precisely on schedule and in the proper orbital position.

Here's a comprehensive look into the science and tactics underlying fine-tuning the trajectory:

The Challenges of Deep Space Navigation:

Gravitational Influences: Multiple celestial entities, including Earth, the Moon, and the Sun exert gravitational forces on Orion along its voyage. These gravitational impacts, however modest, can lead the spacecraft's trajectory to diverge from its planned direction over time.

Launch Imperfections: Even the most painstakingly planned launch may not perfectly inject Orion onto its designated lunar transfer orbit. Tiny mistakes in launch velocity or direction can translate into significant variances over the duration of the multi-day mission.

The Importance of Precision:

Lunar Orbit Insertion (LOI): A successful Lunar Orbit Insertion (LOI) maneuver is vital for mission success. This burn requires Orion to be at the proper location and velocity relative to the Moon for capture into a stable lunar orbit. Deviations from the anticipated trajectory could complicate or even threaten the LOI burn.

Fuel Efficiency: Staying on course eliminates the need for extra engine burns. Excessive propellant expenditure during trajectory correction decreases the amount of fuel available for other essential operations like the LOI burn or potential orbit corrections around the Moon.

The Tools of the Trade: Celestial Navigation and Ground Tracking

Celestial Navigation: This time-tested technique involves using the positions and movements of celestial bodies as reference points. By analyzing the angles between stars or planets, the crew may determine Orion's position and compare it to the anticipated route.

Ground Tracking Stations: A network of powerful ground-based tracking stations on Earth regularly analyzes Orion's signal. These stations measure the Doppler shift of the radio signal transmitted from Orion, which provides information about the spacecraft's velocity relative to Earth.

By integrating data from celestial navigation and ground tracking, mission control can correctly determine Orion's position and course in real-time.

Making Adjustments: Mid-Course Maneuvers (MCMs)

If deviations from the planned trajectory are identified, the crew and mission control will coordinate on a course correction procedure. This correction usually includes making a Mid-Course Maneuver (MCM).

Planning the MCM: Engineers on the ground assess the trajectory deviation and compute the ideal length and direction for a engine burn to correct the route. They factor on fuel efficiency and the intended orbital parameters around the Moon for the LOI burn.

Executing the MCM: The crew receives the approved MCM plan from mission control. They then launch a brief fire of the Orion service module engine in the calculated direction and duration. The engine burn provides the necessary change in velocity to move Orion back into its intended path.

The Art of Efficiency: Minimizing MCMs

While MCMs are required for trajectory adjustment, the idea is to decrease their number to conserve fuel. Mission planners seek to build a route that accommodates for small gravitational factors and achieves the desired lunar orbit insertion point with minimal mid-course burns.

route Optimization Software: Sophisticated software programs play a significant role in planning the most fuel-efficient route. These programs consider in the launch parameters, gravitational forces, and desired orbital features around the Moon to calculate the most efficient path.

Staged Burns: In some situations, a single massive MCM might be broken down into numerous smaller burns spread out over the course of the travel. This piecemeal technique allows for more precise course adjustment while lowering the total quantity of propellant wasted.

Automation and Crew Involvement:

The process of trajectory fine-tuning demands a precise balance between automation and crew input.

Automated Monitoring: Onboard navigation systems continuously monitor Orion's position and trajectory compared to the planned course. Deviations are automatically recognized and reported to mission control for study.

Crew Expertise: The crew receives instruction in celestial navigation and basic orbital mechanics. They can assess trajectory data and provide input to mission control during important decision-making processes linked to course correction actions.

Mission Control Decisions: Ultimately, the decision to perform a MCM rests with mission control. Flight directors and navigation experts assess all available data from the onboard equipment, ground tracking stations, and the crew before issuing directions for a course correction burn.

The Importance of Flexibility:

While the trajectory is meticulously calculated before launch, unforeseen situations could happen during the flight. Space weather events or small equipment problems could necessitate on-the-fly modifications to the trajectory.

Contingency Plans: Mission planners construct contingency plans to manage probable problems that could impact the trajectory. These plans explain potential course correction procedures depending on the nature of the issue discovered.

Crew Training for Adaptability: The crew undergoes considerable training in adapting to unforeseen conditions. They practice responding to simulated trajectory changes and executing course correction maneuvers based on limited information or shifting priorities.

The Reward: A Precise Lunar Arrival

Through a combination of rigorous planning, modern technology, and the skill of the crew and mission control, trajectory fine-tuning ensures Orion's precise arrival at the Moon. This precision opens the path for a successful Lunar Orbit Insertion and sets the foundation for the historic lunar landing and exploration to come. The successful fine-tuning of the trajectory is the culmination of years of engineering effort and human inventiveness, marking a crucial milestone in humanity's return to the lunar surface.

Chapter 4

Lunar Exploration

The Moon, Earth's solitary natural satellite, has captivated humanity for millennia. For millennia, it has served as an inspiration for poets, a beacon for astronomers, and a cosmic dancing partner for our globe. But in recent decades, our obsession with the Moon has moved beyond casual observation to scientific inquiry. The era of Lunar Exploration has come, and with it, the potential of deciphering the mysteries that our lunar neighbor contains.

The Allure of Lunar Exploration:

There are several convincing reasons why the Moon is such a prime target for scientific exploration:

Proximity: The Moon is our closest cosmic neighbor, a modest 238,855 miles (384,400 kilometers) distant at its farthest point. This relative proximity makes it substantially easier and faster to reach compared to other celestial bodies in our solar system.

parallels and Differences: The Moon shares several geological parallels with Earth, such as the presence of craters, volcanoes, and maria (solidified basaltic lava plains). However, it also possesses distinct differences, lacking a global magnetic field and atmosphere. Studying these parallels and variances can provide vital insights into the genesis and evolution of both the Moon and Earth.

Potential Resources: The Moon may have significant resources that could be advantageous for future space exploration initiatives. These resources include water ice locked in permanently shadowed craters, prospective quantities of helium-3 for nuclear fusion, and minerals like iron, titanium, and aluminum.

A Stepping Stone: Lunar exploration acts as a stepping stone for future journeys to Mars and beyond. The Moon provides a platform to test technology, train personnel for deep space missions, and build a prospective base for further space exploration.

Landing on the Moon: A Giant Leap for Mankind

The first manned landing on the Moon took occurred in 1969 as part of NASA's Apollo 11 mission. Astronauts Neil Armstrong and Buzz Aldrin became the first people to walk on the lunar surface, a moment engraved in history as "one giant leap for mankind." Several further crewed missions followed during the 1970s, bringing back rare lunar samples for scientific analysis.

Modern Lunar Exploration: A Global Endeavor

Since the Apollo missions, lunar exploration has become a global undertaking. Several countries, including the United States, China, India, Japan, and the European Space Agency, have active lunar exploration initiatives. These initiatives involve robotic missions that orbit the Moon, land on its surface, and collect data and samples for analysis.

Unveiling the Lunar Landscape:

Lunar exploration missions have revealed a remarkable lunar surface molded by billions of years of impact events, volcanic activity, and exposure to the harsh space environment. Here are some of the important features being studied:

Craters: The Moon's surface is dominated by craters, produced by the impact of asteroids and comets over billions of years. Studying these craters provides insights about the Moon's impact history and the evolution of the solar system.

Maria: These black, flat plains are solidified basaltic lava flows from previous volcanic eruptions. Studying the composition of the maria can offer information about the Moon's interior structure and geological history.

Lunar Highlands: The lunar highlands are the bright, hilly parts of the Moon comprised mostly of anorthosite, a rock type rich in plagioclase feldspar. Studying the highlands can reveal insights regarding the Moon's genesis and early differentiation.

Lunar Resources: A Boon for Future Exploration

Lunar research is not just about scientific discovery; it also has the potential to generate rich resources for future space exploration operations. Here are some of the potential resources being investigated:

Water Ice: Evidence supports the presence of water ice trapped in permanently shadowed craters near the Moon's poles. This ice could be an important resource for future lunar communities, providing drinking water, oxygen through electrolysis, and possible fuel for rockets.

Helium-3: Helium-3 is a rare isotope of helium on Earth but may be more plentiful on the Moon. Helium-3 has potential applications in nuclear fusion energy, delivering a clean and potentially unlimited source of electricity.

Minerals: The Moon may contain precious minerals including iron, titanium, and aluminum. These minerals could be used for in-situ building of lunar dwellings and infrastructure, minimizing the need to carry materials from Earth.

Challenges and Risks:

Lunar exploration involves various obstacles and concerns that need to be addressed:

Harsh Environment: The Moon's surface is a harsh environment with severe temperatures, micrometeoroid bombardment, and fatal sun radiation. Technologies need to be developed to safeguard astronauts and equipment from these threats.

Distance and Cost: Traveling to the Moon is expensive and time-consuming. Developing cost-effective and sustainable transportation systems is vital for future lunar exploration initiatives.

Dust: Lunar dust is a thin, abrasive substance that can damage equipment and pose health dangers to astronauts. Strategies for reducing the effects of lunar dust need to be developed.

Establishing a Lunar Presence: A Long-Term Goal

Many space agencies foresee creating a viable human presence on the Moon. This presence could take the shape of research outposts, mining enterprises, or even lunar cities. A persistent lunar presence might give various advantages:

Scientific Benefits: A permanent presence would enable for long-term scientific research of the Moon and its surroundings. Scientists could perform research in astrophysics, geology, and lunar resource utilization.

Economic Potential: Lunar resources might be mined and employed to fund space exploration activities and potentially even be returned to Earth for commercial applications.

Strategic edge: A permanent lunar presence might provide a strategic edge in space exploration and potentially serve as a launching point for missions to Mars and beyond.

The Future of Lunar Exploration:

The future of lunar exploration is bright. With worldwide collaboration and continuous technical developments, we should expect to see substantial progress in the coming decades. This progress could include:

The establishment of a permanent lunar research station.
The development of in-situ resource utilization (ISRU) techniques to extract and exploit lunar resources.
The use of the Moon as a launching pad for missions to deeper space destinations.

Lunar exploration is not simply about scientific discovery or resource acquisition; it's about pushing the boundaries of human knowledge and exploration. By returning to the Moon and establishing a permanent presence, we take a major leap towards a future where mankind is a multi-planetary species.

Goals of Artemis 3

Following the successful uncrewed Artemis 1 and crewed Artemis 2 flights, Artemis 3 sets its sights on a historic achievement: landing the first woman and the first person of color on the Moon in 2026 (scheduled). This massive project includes a myriad of missions, attempting to push the boundaries of human spaceflight and scientific discovery. Here's a breakdown of the primary objectives:

1. Landing the First Woman and Person of Color on the Moon:

Diversity and Inspiration: Artemis 3 prioritizes diversity in space exploration. By landing the first woman and person of color on the Moon, it wants to inspire future generations from all backgrounds to pursue jobs in STEM disciplines and space exploration.

Expanding depiction: This feat shatters prior limitations and broadens the depiction of humanity in space travel. It conveys a tremendous message of inclusivity and opportunity in the future of space ventures.

2. Establishing a Sustainable Human Presence on the Lunar Surface:

Laying the Foundation: Artemis 3 serves as a critical stepping stone towards a long-term human presence on the Moon. The mission tests the technology and procedures needed for future lunar exploration and potential base development.

Exploration and Resource Utilization: A longer human presence enables for in-depth scientific investigations and exploration of the lunar surface. It also lays the door for potential resource usage, including water ice extraction, for future missions.

3. Demonstrating Advanced Spaceflight Capabilities:

Starship HLS Integration: Artemis 3 marks the first crewed mission utilizing SpaceX's Starship Human Landing System (HLS). This integration highlights the possibility of a reusable lunar lander, a critical technology for future lunar exploration.

Advanced Spacesuits and Systems: Astronauts will employ next-generation spacesuits built for better mobility and functionality on the lunar surface. The mission tests these breakthroughs, pushing the bounds of spacewalking technology.

4. Scientific Discovery and Sample Collection:

The Lunar South Pole: Artemis 3 targets a specific lunar location - the South Pole. This area holds promise for the presence of water ice deposits, a major resource for future exploration. The expedition seeks to collect geological samples from this location for scientific research.

Expanding our Knowledge: Scientific investigations done during the mission will contribute to our understanding of the Moon's geological history, composition, and prospective resources. This knowledge is vital for planning future lunar exploration attempts.

5. A Stepping Stone for Mars and Beyond:

Testing Technologies: The Artemis program, including Artemis 3, acts as a proving ground for technologies and procedures necessary for crewed trips to Mars. Lessons acquired and expertise gained will be crucial for future deep space exploration.

Building Infrastructure: A persistent human presence on the Moon can provide a base for future missions to Mars and beyond. The Moon could serve as a potential launchpad or refueling site for deeper space exploration.

The Artemis 3 mission is a complicated and ambitious project, conveying the hopes for a new era of lunar exploration. Its accomplishment will constitute a tremendous leap forward in human spaceflight, laying the path for a future where mankind has a permanent presence on and beyond the Moon.

Landing Site Selection

The Artemis 3 mission, intending to place the first woman and the first person of color on the Moon in 2026 (scheduled), rests on a vital decision — picking the landing site. This selection procedure entails methodically considering scientific significance, operational practicality, and the capabilities of the Starship Human Landing System (HLS).

Here's a detailed look into the elements affecting the landing location decision for Artemis 3:

The Lunar South Pole: A Region of Interest

NASA has moved its focus to the lunar south pole as the key location of interest for Artemis 3. This location provides various benefits over the equatorial landing sites evaluated during the Apollo missions:

Presence of Water Ice: Evidence shows the presence of water ice trapped in permanently shadowed craters near the poles. Water ice is a crucial resource for future lunar colonies, providing drinking water, oxygen through electrolysis, and potential rocket propellant.

Enhanced Lighting Conditions: Unlike the equator that endures long lunar days and nights, the south pole receives practically continual sunlight for extended periods. This gives more favorable illumination conditions for landing and surface operations during the mission.

Geologically Diverse Terrain: The south pole has a range of geological features, including craters, maria (solidified basaltic lava plains), and hilly regions. This geological diversity affords scientists a lot of material to research the Moon's genesis and evolution.

Candidate Landing Sites: Balancing Science and Safety

Within the broad vicinity of the lunar south pole, NASA has identified 13 particular prospective landing locations. These sites are carefully picked based on numerous criteria:

Scientific Value: The site should offer access to geologically fascinating features and possible resources such water ice. This permits astronauts to collect crucial samples for scientific examination.

Lighting Conditions: The landing site should have sufficient sunlight over the mission period to support solar power generation and give decent vision for landing and surface operations.

Terrain Slope: The optimum landing site should have a generally level and smooth terrain to minimize the dangers associated with touchdown and human mobility during extravehicular activities (EVAs).

Proximity to Resources: Landing closer to prospective resources like water ice minimizes the travel distance for astronauts during surface exploration, preserving time and resources.

Starship HLS Capabilities: The landing site needs to be compatible with the landing and ascension capabilities of the Starship HLS. Factors like topographical risks and illumination conditions need to be considered within the operational limits of the lander.

The Selection Process: Collaboration and Analysis

Choosing the final landing location is a collaborative effort including NASA scientists, engineers, and mission planners. The method involves multiple steps:

Data Analysis: Detailed data from lunar orbiters and landers is collected and examined. This data offers information on the geography, lighting conditions, and potential presence of resources at each candidate site.

Risk Assessment: Each candidate site is analyzed for potential dangers such craters, slopes, etc Each proposed site is analyzed for potential risks including craters, slopes, and loose regolith (lunar soil). The purpose is to minimize risks connected with landing, astronaut mobility during extravehicular activities (EVAs), and potential equipment damage.

Simulations and Modeling: Engineers employ complex computer models and simulations to analyze the feasibility of landing the Starship HLS on each prospective site. These models take into account elements including terrain slope, illumination conditions, and the lander's propulsive capabilities.

Public Input: In some situations, NASA may invite public input on the landing site choices. This can assist extend scientific involvement and inspire public interest in the mission.

Making the Final Choice:

After considerable investigation, risk assessment, and teamwork, a definitive landing site is determined for Artemis 3. This decision includes all the aforementioned variables, establishing a balance between scientific relevance, operational viability, and crew safety.

The Importance of Landing Site Selection:

The choice of landing site has a considerable impact on the success and scientific outcomes of Artemis 3. A well-chosen site can offer astronauts with access to significant scientific data and resources, while also assuring a safe and successful landing and mission.

The Future of Lunar Exploration:

The landing site selected for Artemis 3 sets the scene for future lunar exploration attempts. The data obtained and the knowledge gained during this mission will be vital for determining future landing site decisions as we build a more permanent human presence on the Moon. This persistent presence offers the potential for uncovering more of the Moon's mysteries and paving the road for future trips to Mars and beyond.

Surface Activities and Scientific Objectives

The Artemis 3 mission, intended to land the first woman and the first person of color on the Moon in 2026 (planned), promises to represent a historic leap forward in lunar exploration. Beyond the historic firsts, the expedition is replete with scientific fascination. The crew will execute a range of surface activities aimed to meet a multitude of scientific objectives, pushing the boundaries of our understanding of the Moon. Let's go into the scheduled surface activities and the scientific knowledge they intend to unearth.

Surface Activities: A Week on the Lunar Frontier

The Artemis 3 astronauts will spend a key week on the lunar surface, undertaking a variety of planned activities:

Moonwalks (EVAs): Multiple Extravehicular Activities (EVAs) are scheduled, averaging roughly 4-8 hours each. During these moonwalks, astronauts will go out of the lander to explore the surrounding lunar surface and perform scientific studies.

Sample Collection: A key priority will be gathering a diverse range of lunar samples. These samples will comprise rocks, regolith (lunar dirt), and maybe ice cores from permanently shadowed craters. The samples will be meticulously documented and returned to Earth for in-depth analysis.

Deploying Scientific devices: The crew will deploy a range of scientific devices on the lunar surface. These instruments may include seismometers to examine lunar earthquakes, magnetometers to detect the Moon's magnetic field, and instruments to evaluate the composition of the lunar surface.

Maintaining the Lander: The astronauts will execute crucial maintenance chores on the Starship HLS lander to enable its safe ascension from the lunar surface at the end of the mission.

Scientific Objectives: Unveiling a Lunar Treasure Trove

The surface activities during Artemis 3 are methodically organized to address a range of scientific objectives:

Understanding the Moon's Formation and Evolution: Studying lunar rock and soil samples will reveal insights into the Moon's formation process, its composition, and its geological history. This can tell us more about the development of our solar system and the early Earth.

Searching for Evidence of Water Ice: A significant objective is to study the occurrence of water ice deposits, particularly near the lunar south pole (the targeted landing zone). The crew will gather ice core samples to validate the presence, abundance, and chemistry of aqueous ice on the Moon.

Exploring Lunar Resources: Artemis 3 intends to explore the potential of exploiting lunar resources. By examining the composition of the lunar surface, scientists can learn more about the availability of resources including minerals, metals, and possible helium-3 for future fusion energy applications.

Testing Technologies for Future Exploration: The mission serves as a testing ground for technologies critical for future lunar exploration and eventual crewed missions to Mars. This includes testing new spacesuits, tools, and research instruments in a true lunar environment.

Characterizing the Lunar Environment: By deploying scientific instruments and conducting experiments, scientists aim to gain a better understanding of the lunar environment, including the radiation levels, the presence of a tenuous lunar atmosphere, and the impact of micrometeoroid bombardment on the lunar surface.

A Collaborative Endeavor: Science and Exploration Hand-in-Hand

The scientific objectives of Artemis 3 are determined through collaboration between NASA scientists, astronauts, and foreign partners. This collaboration guarantees that the mission tackles the most urgent concerns concerning the Moon and maximizes the scientific return on investment.

The Legacy of Artemis 3: A Stepping Stone to the Future

The surface operations and scientific aims of Artemis 3 are not just about comprehending the Moon. They constitute a stepping stone for future space exploration operations. The knowledge obtained and the technologies tested during this mission will pave the road for establishing a sustainable human presence on the Moon and ultimately, propel humanity into Mars and beyond.

By uncovering the secrets buried within the lunar surface, Artemis 3 promises to usher in a new age of scientific discovery and human exploration in our solar system.

Chapter 5

Crew Operations and Equipment

The Artemis 3 mission, aimed for a lunar landing in 2026 (scheduled), is a huge breakthrough in human spaceflight. This historic mission will be carried out by a painstakingly chosen crew, each person possessing a distinct skillset and essential responsibilities.

Crew Composition:

The actual number of crew members on Artemis 3 has still to be finalized, although it is believed to be around four astronauts. Here's a summary of the potential crew roles:

Mission Commander: An experienced astronaut responsible for the overall leadership and decision-making during the mission. They will monitor all crew actions, interact with mission control, and assure the success of the mission objectives.

Pilot: Another highly experienced astronaut responsible for operating the Orion spacecraft during several mission phases, including launch, Earth orbit maneuvers, lunar orbit insertion, and re-entry.

Mission Specialist 1 (MS-1): An astronaut with specialization in geology, planetary science, or a similar discipline. MS-1 will be significantly involved in conducting scientific studies on the lunar surface, collecting samples, and operating scientific instruments.

Mission Specialist 2 (MS-2): An astronaut with competence in engineering or space systems. MS-2 will be responsible for running the Starship HLS during lunar descent and ascension, completing maintenance chores on both Orion and the lander, and providing technical assistance during the mission.

Emphasis on Diversity and Experience:

The selection of the Artemis 3 crew will stress diversity and experience. This means including astronauts from varied backgrounds and countries, guaranteeing a well-rounded team capable of managing the challenges of lunar exploration. Furthermore, the crew will undertake comprehensive training to prepare them for all areas of the mission, including spaceflight protocols, scientific operations, geological fieldwork, and emergency response circumstances.

Vehicles and Tools for Lunar Exploration: A Synergistic Approach

The Artemis 3 project relies on a potent combination of spacecraft and lunar surface technology to achieve its goals. Here's a deeper look at the essential trucks and tools the crew will utilize:

Orion Spacecraft: The Orion multi-purpose crew vehicle will act as the primary transporter for the Artemis 3 crew. Orion will launch the astronauts from Earth, sustain them for the journey to the Moon and return, and provide a safe environment for re-entry into Earth's atmosphere.

Starship Human Landing System (HLS): Developed by SpaceX, the Starship HLS will be responsible for landing the crew on the lunar surface and launching them back into lunar orbit for rendezvous with Orion. This reusable lander marks a significant leap in lunar landing technology.

Lunar Rover: The specific lunar rover model for Artemis 3 is still to be chosen. However, the crew will utilize a pressurized rover capable of moving personnel and equipment across the lunar surface during their EVAs (Extravehicular Activities). The rover will extend the crew's reach and allow them to explore a greater geographical area during their short stay on the Moon.

Scientific Equipment: The crew will be outfitted with a number of scientific tools to perform study on the lunar surface. These instruments may include tools for collecting rock and soil samples, seismometers to research moonquakes, magnetometers to detect the Moon's magnetic field, and spectrometers to evaluate the composition of the lunar surface.

Mobility and Life Support: The crew will wear next-generation spacesuits specifically suited for lunar exploration. These spacesuits will offer better mobility and flexibility compared to earlier models, allowing astronauts to conduct a larger range of tasks during their moonwalks. Additionally, a portable life support system will give the crew with oxygen, water, and thermal regulation during their EVAs.

Lunar Rover and Equipment Deployment: Extending Reach and Scientific Inquiry

The lunar rover serves a critical role in maximizing the crew's scientific return during

Artemis 3. Here's how the rover and equipment deployment will be conducted:

Offloading the Rover and Equipment: Upon landing on the Moon, the crew will utilize the Starship HLS's cargo hold to offload the lunar rover and numerous scientific equipment. This technique will likely be helped by the Canadarm3, a robotic arm on the Orion spacecraft, enabling safe and efficient deployment.

Setting Up the Lunar base: Once the rover and equipment are on the surface, the team will create a temporary lunar base. This may require unfolding livable modules, deploying solar panels for power generation, and setting up communication relays to maintain contact with Earth.

Utilizing the Rover for Exploration: The lunar rover serves as the crew's primary means of mobility throughout their surface excursions. They will utilize the rover to travel to several geological places of interest, gathering samples, deploying scientific instruments, and conducting geological surveys. The greater range provided by the rover allows the crew to explore a wider lunar area and acquire more scientific data.

Recovering Samples and Equipment: At the conclusion of their lunar stay, the crew will carefully collect all the scientific samples and equipment back onboard the Starship HLS for their return journey. This rigorous approach ensures that the essential scientific data and findings from the lunar surface are preserved for study back on Earth.

A Synergistic Mission Design:

The combination of Orion, Starship HLS, the lunar rover, and a suite of scientific instruments forms a synergistic mission architecture for Artemis 3. Each aspect serves a key role in assisting the crew and completing the mission's scientific objectives. The successful installation of these vehicles and instruments will unveil a wealth of knowledge about the Moon and pave the road for future human exploration of our solar system.

Scientific Instruments and Discoveries

The Artemis 3 mission, slated for a lunar landing in 2026, offers tremendous scientific significance. Beyond the historic human achievement, the mission intends to unlock fresh discoveries about the Moon with a set of scientific instruments deployed by the crew. Let's dig into the intended scientific payloads and the potential discoveries they might facilitate.

Overview of Scientific Payloads: A Diverse Arsenal for Lunar Exploration

The specific scientific equipment chosen for Artemis 3 will be finalized after a rigorous selection procedure. However, we should expect a broad selection of payloads designed to address distinct scientific aims. Here are some prospective devices and their areas of investigation:

Geochemical Analyzers: These equipment will be utilized to evaluate the composition of lunar rocks and soil samples gathered by the crew. This research can disclose information on the Moon's geological history, the presence of specific minerals or resources, and the processes that created the lunar surface over billions of years.

Seismometers: Deploying seismometers on the lunar surface will allow scientists to examine moonquakes, also known as lunar seismic occurrences. By evaluating the frequency and location of these events, we can acquire insights regarding the Moon's internal structure, the presence of a molten core, and the ongoing geological activity within the lunar interior.

Magnetometers: Measuring the Moon's magnetic field, albeit considerably weaker than Earth's, can reveal vital information about the Moon's history and interaction with the solar wind. These data can give light on the Moon's previous magnetic activity and its role in safeguarding the lunar surface from radiation.

Remote Sensing Instruments: While some instruments require direct connection with the lunar surface, others may work from a distance. Remote sensing devices like lidar or radar can be used to map the lunar surface topography in great detail, identify suitable landing locations for future missions, and look for evidence of water ice deposits hidden in permanently shadowed craters.

These are only a few examples, and the final payload will likely comprise a broader range of sensors chosen based on specific scientific priorities and technology breakthroughs at the time of launch.

Highlighting Specific Instruments and Potential Discoveries: A Look at Three Key Payloads

NASA has already picked the first set of instruments for deployment by the Artemis 3 crew. Let's study these devices and the potential discoveries they might facilitate:

1. Lunar Environment Monitoring Station (LEMS): Developed by JAXA (Japan Aerospace Exploration Agency), LEMS is a set of instruments designed to research the lunar seismic activity and the lunar environment. LEMS could help us comprehend the frequency and intensity of moonquakes, revealing insights about the Moon's interior structure and potential geological processes occurring beneath the surface.

2. Lunar Effects on Agricultural Flora (LEAF): This unique payload, designed by Space Lab Technologies, aims to examine how plants respond to the lunar environment. By growing small plant samples on the Moon, LEAF could provide vital insights for building technology for future lunar agriculture and potential food production during extended lunar missions.

3. Lunar Dielectric Analyzer (LDA)

This Japanese equipment built by the University of Tokyo will measure the electrical properties of the lunar regolith (soil). Here's how it could lead to amazing discoveries:

Water Ice Detection: The LDA can detect fluctuations in the electrical characteristics of the regolith, perhaps suggesting the presence of water ice deposits buried beneath the lunar surface. This information is vital for proving the existence and abundance of water ice on the Moon, a potential resource for future lunar communities.

Subsurface Layering: By evaluating the electrical properties at different depths, the LDA can assist map the subsurface layering of the Moon. This information provides insights into the Moon's geological history and the distribution of different components inside the lunar regolith.

Resource Exploration: The LDA may also be used to identify the presence of other resources like minerals or metals by detecting differences in their electrical conductivity. This information can be important for appraising the possibilities for future in-situ resource utilization (ISRU) on the Moon.

Beyond the Expected: The Potential for Unforeseen Discoveries

The joy of scientific exploration resides in the prospect of unforeseen findings. The equipment deployed during Artemis 3 might disclose phenomena or clues that we don't even expect now. These unexpected findings could change our knowledge of the Moon and the solar system as a whole.

A Stepping Stone to the Future: The Legacy of Artemis 3 Scientific Discoveries

The scientific discoveries achieved during Artemis 3 will have a enduring impact on our understanding of the Moon. The data acquired will not only solve our current questions but also pave the road for future lunar exploration attempts. This understanding will be important for creating a sustainable human presence on the Moon and ultimately, moving humanity into exploring Mars and beyond.

The Artemis 3 mission represents a huge leap for science and exploration. By deploying a sophisticated set of scientific tools, we stand on the cusp of uncovering the Moon's secrets and launching a new era of discovery in our cosmic neighborhood

Challenges and Delays

The Artemis program, seeking to return humans to the Moon by 2026, has a range of problems that can cause delays and blockages. Here's a closer look at the development difficulties, the reasons behind mission delays, and the future prospects for this massive lunar exploration endeavor.

Development Challenges and Solutions: A Race Against Technical Hurdles

Starship Development: The Starship heavy-lift launch vehicle built by SpaceX is a vital component for Artemis 3. However, Starship is still under construction, and early test flights have met technical concerns like vehicle explosions and engine failures. These issues demand careful testing and refinement before human grading can be attained.

Spacesuit Redesign: The next-generation spacesuits developed for Artemis moonwalks have encountered criticism regarding mobility limits and possibly thermal control difficulties. Engineers are working on modifying these suits to offer better flexibility and improved life support systems for astronauts during lunar EVAs (Extravehicular Activities).

Lunar Lander Integration: Integrating the Starship HLS (Human Landing System) with Orion to allow safe docking and crew transfer in lunar orbit offers a huge engineering challenge. This integration requires comprehensive testing and verification to guarantee crew safety during crucial mission phases.

Overcoming these difficulties involves:

Rigorous Testing: Implementing intensive ground testing and flight qualification programs for both Starship and the integrated Orion-Starship system is required.

International Collaboration: Leveraging the knowledge and resources of international partners like the European Space Agency (ESA) and the Canadian Space Agency (CSA) can speed problem-solving and technical improvement.

Prioritizing Safety: Maintaining a uncompromising focus on crew safety should lead all decisions relating to testing, development, and mission planning.

Timeline Adjustments and Mission Delays: A Matter of Prudence

The Artemis program has faced multiple timeline modifications due to the development issues outlined above. Here's a summary of the reasons behind these delays:

Technical Uncertainties: Unforeseen challenges identified during Starship development and spacesuit testing needed additional time for resolution. Rushing these processes could compromise mission success and crew safety.

Budgetary Constraints: Securing sufficient funds for the full Artemis program remains a challenge. Budgetary limits can effect the pace of development and testing activities.

Prioritization of Safety: The decision to delay missions prioritizes comprehensive testing and risk minimization. This technique offers a higher possibility of mission success and the safety of the Artemis astronauts.

Future Prospects for Artemis Program: A Journey of Perseverance

Despite the hurdles and delays, the future of the Artemis program remains bright. Here's a preview of what is ahead:

Gradual evolution: We should expect a gradual evolution of missions, commencing with uncrewed Artemis flights to test systems extensively before committing to crewed landings.

Sustainable Lunar Presence: The ultimate goal of Artemis is to establish a sustainable human presence on the Moon. This will require creating lunar habitats and developing technology for in-situ resource utilization (ISRU).

Gateway development: A vital part for future lunar exploration is the development of the

Lunar Gateway, a mini-space station circling the Moon. The Gateway will function as a staging site for astronauts and a communications center for lunar surface operations.

International Collaboration: The Artemis program encourages international collaboration. Partnerships with space agencies worldwide will bring together knowledge, resources, and a global vision for lunar exploration.

Inspiration for the Next Generation: The Artemis missions contain great potential to inspire the next generation of scientists, engineers, and explorers. The initiative can kindle a love for STEM education and encourage a new wave of innovators to push the boundaries of space exploration.

A Stepping Stone to Mars: The Artemis Legacy

The Artemis program is a vital stepping stone on the way towards human exploration of Mars. The hurdles conquered during Artemis missions will provide essential knowledge and technology breakthroughs necessary for future crewed missions to the Red Planet.

By building a lunar foothold, we may test technology for long-duration space travel, create sustainable life support systems, and gain operational expertise in a faraway celestial body. The Artemis program prepares the stage for humanity's next major leap in space exploration, moving us towards Mars and beyond.

The road to Artemis 3 is riddled with hurdles, but the potential benefits are great. With perseverance, worldwide collaboration, and a commitment to safety, the Artemis program holds the prospect of exposing the Moon's secrets and ushering in a new era of human discovery in our solar system.

Conclusion

As you close this book, the lunar dust settles on the pages, a silent reflection of the enormous adventure that is Artemis 3. The challenges we've examined may appear formidable, the delays aggravating. But remember, humanity has always thrived on delving into the unknown, on pushing the frontiers of what's possible. The Moon, long a celestial dream, now hangs within our grasp.

The achievements of Artemis 3 will resonate far beyond the lunar surface. The scientific discoveries will renew our awe of the cosmos and change textbooks. The technological advances will spill over, altering our existence on Earth in ways we can scarcely comprehend. But perhaps the most deep impact resides in the inspiration it generates.

Imagine a little girl, staring at the Moon with renewed amazement, her imagination overflowing with possibilities. Maybe she'll be the engineer who designs the next generation of spaceships, the scientist who uncovers the secrets of lunar ice, or the astronaut who takes the first human steps on Mars. The Artemis program is a seed put in the rich ground of human ambition, and its fruits will be reaped for years to come.

So, once you turn the final page, don't perceive it as a conclusion, but as a wonderful stop. The rockets will roar again, the brave crew will embark, and humanity's adventure on the Moon will unfold. This is not only about a voyage to the Moon; it's about a great stride towards a future where mankind reaches for the stars, not as a distant dream, but as a birthright waiting to be taken.

www.ingramcontent.com/pod-product-compliance
Lightning Source LLC
Chambersburg PA
CBHW071058240526
45471CB00016B/1990